Yang Yin Theory

By Dennis L. Muldrow

ISBN: 0615895263
ISBN-13: 9780615895260

DEDICATION

I dedicate this book to my Father who always asked "What would happen when an unstoppable force met an unmovable object?"

iv

CONTENTS

ACKNOWLEDGMENTS

I would like to thank all the scientists throughout history and those to follow for their dedication in the quest for truth. Without such people and their dedication our species would still be dwelling in caves. They have and will pave the way to a Theory of Everything.

Forward

If a theory makes us mere humans out to be more than what we are, there is a high probability the theory is wrong.

Starting with a single singularity before time and space is an egocentric ideology and falls in the same realms as the earth being in the center of the universe.

The Yang Yin Theory is based on the premise infinity is a reality. Infinity can be the only true starting point for anything existing before time and space.

The Big Bang Theory is the dominant theory of our era. It is clearly not a theory of everything. The Big Bang theory is more of a working theory in progress.

The starting point for all things created is crucial. If the foundation is not true all else that follows will need extra work exponentially to try to correct the mistakes. The work that is needed in the Big Bang theory is all the unsolved problems in physics left in its wake.

The answer to the four fundamental forces should validate any theory on the origin of the universe and is the pathway to a theory of everything. No answers, no validation, go back to square one and find the mistakes.

Square one in this case, is a singularity preexisting before time and space. The mistake is not acknowledging the collapsing force that is needed to preexist before time and space. Not only is this collapsing force needed to create a big bang event it has to be an accumulative force to create the energy needed to create what happens during and what happens after a big bang event. A single collapsing force is not enough.

This is a huge mistake the Yang Yin Theory will rectify. The preexisting collapsing force not accounted for is the force behind all other forces.

Once explained the viewer will find reality is a far more exciting place. Our reality, our bubble of time is a great balancing act in motion.

The Yang Yin Theory will show why infinity is the most probable starting point and explain its structure and the preexisting laws of physics that govern it.

Science is a slow moving process, which is the price that must be paid in the quest for truth. It is better than trying to push the same old story without regard to the truth.

 Come join the next revolution in science "The Yang Yin Theory"

1

INFINITY

Infinity is an entity that has no beginning, no end, no center, no edge, and has always been and always will be.

Science does not like the concept of infinity because it cannot be demonstrated therefore it cannot be proven. The law of conservation, all that exists in this universe can never be completely destroyed only altered, implies infinity.

The law of conservation means if this universe were created from an infinite state, it would not be in an empty state, it can only be an alteration of what now exists.

There are too many endless possibilities void of structure for creation to come from infinity of empty space. In the realms where all things are possible the laws of physics become meaningless.

This universe comes from a structured past, from a system able to build upon itself and evolve over a period of time. This is self, evident in most everything that exists in this universe, Sub atomic particles, atoms, molecules, life.

The first law of infinity is to remain infinite or it would not be infinity.

The second law of infinity is all finite entities existing in infinity will become infinitely smaller. For infinity to exist all points leading away from a finite entity the finite entity must get infinitely smaller in all directions. The second law of infinity is one reason a preexisting collapsing force exists.

A finite entity cannot collapse into itself infinitively. It will eventually run into itself and expand outwardly in all directions. This is another preexisting law that must exist for a big bang event to take place.

The question is how is infinity structured, an infinite state capable of evolving over a period of time from a system able to build upon itself?

A single singularity preexisting before time and space does not answer this question. It contains too much energy. If it can preexist once without explanation it can preexist again and again in the same arena. All the mass and energy that created this universe would big bang in this universe over and over. The effect of this would be observable.

The form of structures created from an infinitive state is most likely a form an infinitive state is in.

In this universe there are a few basic patterns. These patterns are single three dimensional spheres (stars, planets, moons, black holes, structure of atoms as a whole.)

There are connected spheres (molecules, nucleus of atoms, quarks.)

There are cork screw events. This includes all rotations, orbits, and spins moving through the arrow of time and wavelengths.

There are two universal forces, Gravity and the expansion force of the universe.

In the creation of a universe, the correct pattern is not a single sphere that would need an explanation of its singled out existence. The correct pattern is infinitely connected spheres which existence can be explained.

All that exists in this universe cannot be completely destroyed only altered. The question is what is the infinite unaltered state? I call it expanded quantum macro foam (EQMF). Planck constant states a physical action cannot take on any indiscriminate value. This value relates to one of the basic patterns in our universe the corkscrew event (wavelength). There is a reason why all things spin and quantify. Time is a three dimensional rotating collapsing sphere one after another. Energy moving against collapsing negative time and dimension will quantify on the inside of each collapsing sphere of time at different latitudes according to the energy's velocity. Energy quantifies by carving out its own moment in time while heading into the next moment of collapsing time each moment larger than the next.

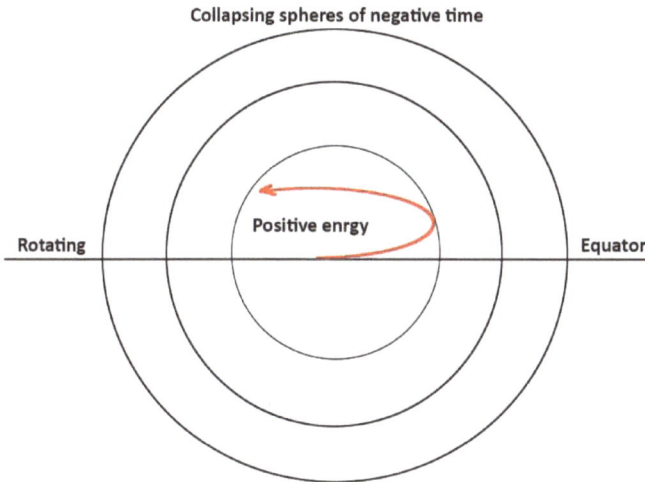

Collapsing spheres of negative time

Rotating

Positive enrgy

Equator

Spinning continuously is like an ice skater spinning and able to pull their arms in infinitively at the same rate as resistance.   The opposite is an internal expansion rate expanding faster than an energy packet's spin stopping its rotation. Planck constant would cease to exist in an infinitive state. Each packet cannot stop its spin and expand infinitely because there can only be one infinite state. Planks constant and reduced constant becomes Planck expanded constant.

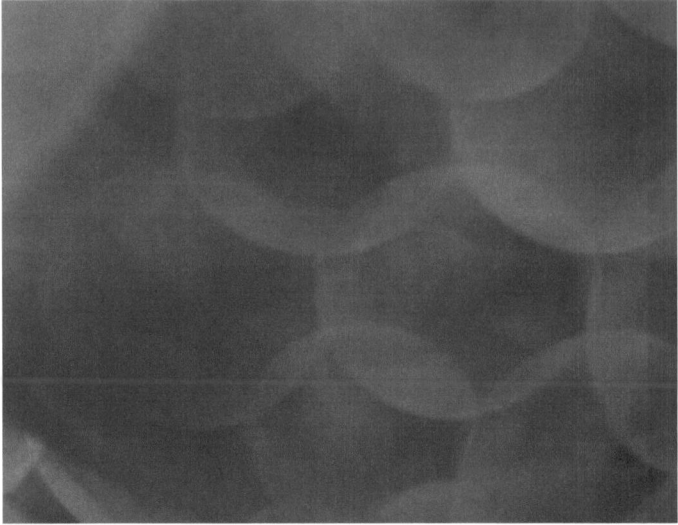

Infinity is infinitely connected three dimensional spheres in a semi steady state. It is in a semi steady state because it has occasional vibrations running along its borders (its strands). This is much like String Theory only all the strands are infinitely connected with infinite dimensions. It cannot be in a complete steady state or nothing would happen. Universes begin when two opposing vibrations run into each other on the same strand and the strand breaks. A finite entity is born.

Infinity will continue to collapse on all finite entities until all finite entities return to an infinite state. Collapse and expansion is part of the process.

Infinity has an infinite amount of universes that have and are going through this process. There are an infinite amount of dormant universes that have been put back into an infinite state.

Beginning of a universe

All that is here, all that exists is merely to mend a tear in infinity. If it can happen once it can happen, an infinite amount of times.

Infinity of an endless sea of expanded quantum macro foam is the starting point for all finite progressions.

Expanded quantum macro foam (EQMF) is everything in this universe expanded to the point where all is reduced to energy and energy is expanded into an infinite state of potential energy.

Once a strand breaks, which holds a sphere together, the finite entity created reduces in size and pulls in the surrounding three dimensional spheres. At a certain point their strands break creating a larger collapsing finite entity.

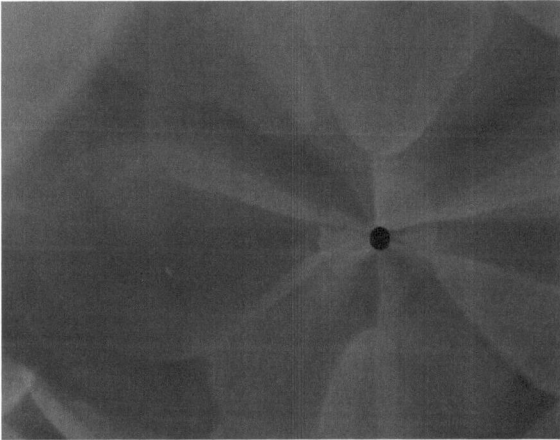

This process will continue each three dimensional collapse one after another. Each three dimensional collapse containing more energy as it pulls on more of the surrounding EQMF of potential energy. Once disconnected, broken three dimensional spheres of potential energy becomes energy.

This creates collapsing spheres one inside another, each new sphere gaining more energy creating a system able to build upon itself. Once started potentially the collapsing force can collapse indefinitely and will not stop until all the energy created is converted back in its infinite state as EQMF. Like a scar in the human skin, once all the energy in a universe is put back into an infinite state it will not resemble the original state from which it came. It becomes jumbled up spheres of quantum macro foam. A dormant universe expanded into an infinite state. When a new universe begins and its collapsing spheres increase in size it can run into and incorporate a dormant universe waking it up like a seed. All the matter in a universe does not need to be created in a single big bang event.

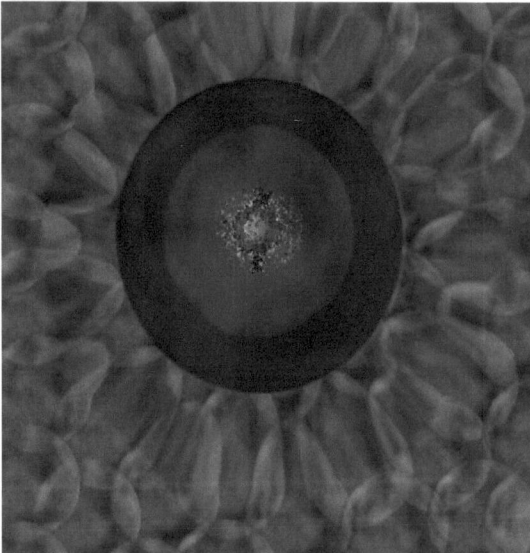

Each collapsing sphere is finite and will eventually collapse into itself unless altered by another force. Once a sphere collapses into itself it will expand outwardly in all directions unless altered by another force. The energy heading outwardly in all direction will run into the next collapsing sphere surrounding it. These two opposing forces are able to break each other up. Differences between these two opposing forces have the potential to create wavelengths, particles, and space.

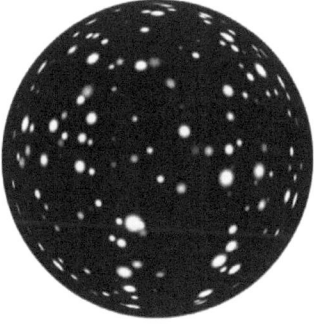

I do not believe in a single big bang event. Like everything else in this universe, it had to evolve into a big bang event. Collapse and expansion are preexisting forces and mean nothing if they are not continually fed. All new particles created in a new universe continuing to collapse would continue to collapse and expand individually breaking up and creating more particles that will go through the same process. The process would be over almost as fast as it started returning back to quantum macro foam leaving a scared area of three dimensional spheres if no other force existed.

A new force is introduced when collapsing negative time, always increasing in size with each new collapse, runs into a dormant universe locked into an infinite state. A dormant universe as it is captured into a new collapsing system, at first, acts like and anchor. This causes the three dimensional collapsing sphere of negative time to spiral and each new collapse as well. This introduces a new force, centrifugal force.

With the introduction of centrifugal force the dynamics of collapse and expand have changed. After a collapse, positive energy expanding against the next collapsing sphere of negative time and dimensions also spirals, taking longer to collapse the positive energy back into itself because it has centrifugal force, its own tensor strength.

The longer it takes to collapse positive energy back into itself creates more space. The more space created increases velocity for the next outwardly expansion. The more energy's velocity increases the more collapses it takes to collapse positive energy into negative time creating more space. This creates away for positive energy to build upon itself as negative energy builds upon itself.

EQMF is an endless sea of connected three dimensional spheres. They are not perfect spheres that would create voids. When they collapse it is a four dimensional collapse (3 dimensions and time) not to be confused with the collapse of the empty space we observe. The empty space we observe is after each three dimensional collapse of negative time is absorbed by all the matter in the universe heading into positive time and dimension. The more mass heading into positive time and dimension the more encapsulations of negative time it will take to pull it into negative time and dimension. The more encapsulations of negative time a mass heading into positive time and dimension acquires the more it is delayed in the expansion rate of positive time and dimension. This delay into positive expanding time and dimension is the reason for gravity.

This system able to build upon itself in the creation of a universe would take less energy than it would to form a human thought.

Time

Time resembles a looped magnetic field.

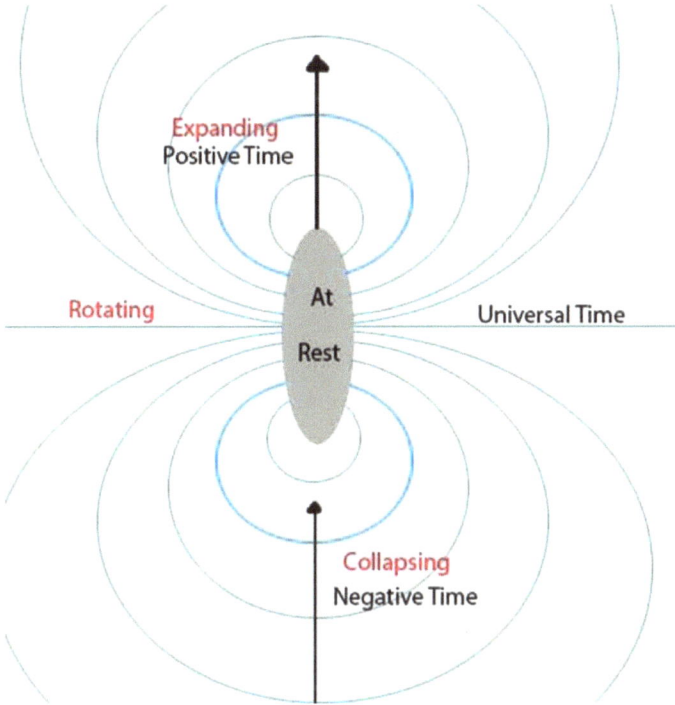

There are different generations as well as different variations of time. There is negative collapsing time, positive expanding time, special relativity, general relativity, and spacetime. The generations are spacetime, electron influenced time, electromagnetic looped time, and Black hole influenced time.

The difference between negative collapsing time and positive expanding time creates spacetime.

Spacetime is created by all the matter in the universe absorbing negative collapsing time and dimensions.

Spacetime increases in volume in different ways.

When negative collapsing time increases in volume and is absorbed spacetime increases in volume, as new matter (dark matter) is introduced into the system matter as a whole in the universe absorbs negative time and dimensions at a faster rate, as matter increases in velocity gaining more positive energy, and as positive expanding time expands into the future.

Spacetime is in a zero expanding dimensionless state giving it three dimensional qualities. Spacetime is in a quasi-infinitive state where different quantum states of time can exist and appear to exist on the same plane.

The rotation of collapsing negative time and dimensions creates the ability for special relativity and the ability to quantify in a moment of time. The equator of rotating negative collapsing time and dimensions creates general relativity (universal time).

The rotation creates centrifugal force. Because it is a dimensional collapsing sphere it collapses at different times at different latitudes because of centrifugal force.

 Energy with enough velocity heading into positive expanding time and dimensions has the ability (starting at the equator of universal time) to circumnavigate along the inside of the next collapsing sphere of negative time and dimensions. This allows positive energy to move into higher latitudes on the inside of a collapsing sphere according to its velocity. Energy will then quantify at the latitude it is in as negative collapsing time and dimensions collapses and expands into the next moment creating a packet of energy. Mass or matter has the same ability having positive energy moving with the expansion force of the universe. Mass or matter will also quantify at higher latitudes on the inside of a sphere of collapsing negative time and dimensions. Mass or matter is an entanglement of positive expanding time and dimensions and negative collapsing time and dimensions. This gives mass or matter the ability to absorb negative collapsing time and dimensions at a faster rate according to its mass and its positive energy. This reverses the effect of how time collapses around mass or matter. Collapsing negative time and dimension is pulled into matter according to its mass and positive energy. The higher the latitude mass or matter quantifies the lower it is according to universal time giving mass or matter the Gravitational effect on spacetime. Energy in a wavelength packet does not have mass so it does not reverse the effect. Energy packets move into each expanding moment of spacetime at the rate collapsing negative time and dimensions is being absorbed by all the matter in the universe which just happens to be the speed of light. This creates the spinning ice skater effect being able to pull their arms in infinitely. Packets of energy are not reducing in size the empty void of space, its environment is continually getting larger.

Evolution of Matter

Wavelengths

Collapse and expansion is how the evolution of matter begins. The entanglement of these two entities, are observed in different fashions and on different planes. Wavelengths are created first in a newly forming universe. The energy heading into positive expanding time and dimensions has the ability to circumnavigate along the inside of a collapsing sphere of negative time and dimensions as it collapses. Positive energy creates its own accelerated centrifugal force moving against a rotating collapsing sphere of negative time and dimensions. Its centrifugal force accelerates as positive energy increases its latitude in the sphere of negative collapsing time and dimensions. A packet of energy quantifies when its centrifugal force is equal to or greater than that of a rotating sphere of negative time and dimension's force as it collapses in one moment of time.

Wavelengths are massless because they only take one moment in time to create. Wavelengths self-perpetuate using the collapse of negative time and dimensions and the expansion of space. A wavelength acts like an inchworm as it eccentric orbits around itself into the next moment. This is why wavelengths have particle, wave duality. Magnetic moments are caused by a wavelength's centrifugal force. It takes longer for a sphere of collapsing negative time to collapse around the equator of a rotating entity with positive energy. This creates a separation in time as the axis of a rotating entity with positive energy collapses first before the equator.

Each collapse and expansion of a new universe produces stronger wavelengths as they increase in latitude on the inside of each new sphere of collapsing negative time and dimensions. This is why a wavelength's frequency increases with velocity. Wavelengths always start at the equator of collapsing negative time until they quantify. The new universe fills with low frequency waves and progress with the spectrum of wavelengths as latitudes increase with each new collapse and expansion until it reaches the axis with gamma rays.

The influence of collapsing negative time and dimensions

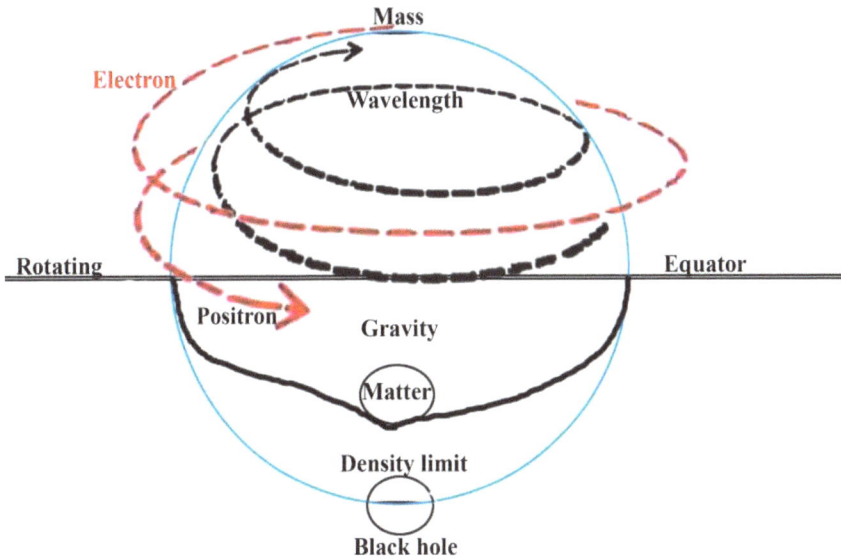

Electron

The first mass to be created in a new universe is not quarks. It is the electron. Electrons have enough positive energy to circumnavigate on the inside of a collapsing sphere of negative time and dimensions as it rotates increasing latitude until it comes out of the axis of negative time and dimensions. Coming out of the axis of collapsing negative time and dimensions an electron's energy is able to circumnavigate along the outside of the sphere of collapsing negative time and dimensions as it collapses. This gives an electron the ability to take up more than one moment of time. The entanglement of both opposing forces taking more than one moment of time creates mass.

An electron has the ability to filter time through it at different rates. Circumnavigating around the outside of a rotating sphere of collapsing negative time and dimensions creates the ability to filter time and energy. When energy is exerted upon or taken away from an electron its energy moves up or down the latitude of the rotating sphere of collapsing negative time and dimensions on the outside. More energy the more it moves towards the equator of a collapsing sphere. Moving towards the equator down from the axis allows an electron to occupy more space and time. Energy moving back towards the axis allows an electron to occupy less space and time. This gives an electron the ability to fall into atomic shells or squeeze out of them. Electrons act like a flywheel in a bicycle gear in an atomic structure.

Moving from the inside of a collapsing sphere of negative time and dimensions then up and around the outside as it collapses gives an electron a second generation collapsing quality. An electron is able to filter time through it by increasing its volume.

The ability to filter time exerted upon it makes an electron a miniature fluctuating wormhole event. This gives an electron the ability to alter time.

To a theorist theorizing about a theory of everything or the origin of the universe the electron is the great validator. An electron is able to combine its energy with other electrons and show in detail the separation of time and space. An electron is able to show the two dominant forces in the universe, the existence of special relativity, the curvature of time and space, collapsing spheres of negative time and dimensions, that our future is a spiraling event.

Magnetism

Magnetism is when most of the electrons in a highly conductive material orbit and spin in zones in relatively the same direction. The spin and orbit are important because it creates centrifugal force which creates special relativity at different latitudes in a rotating sphere of collapsing negative time. This creates a delay in time at the equator and the plane of orbit at high velocities. A magnet proves 95.5 percent the Yang Yin theory is correct. A magnet creates a separation in time which allows its axis to go through the process of time before its equator. This ability reveal the two dominant forces in the universe at the axis which is expanding positive time and dimensions (North Pole), and collapsing negative time and dimensions (South Pole).

An electrons ability to filter energy through it gives it the ability to filter time through it. An electron can speed up or slow down the process of time. The shells in an atomic structure, is a slowed down process of collapsing negative time and dimensions.

Electromagnetic Field

Sending a charge of electricity down a conductive wire produces an electromagnetic field around the wire. A charge of electricity in a conductive wire is produced when electrons in the conductive wire are pushed into filtering time at a faster rate. This makes the electrons in the conductive wire want to expand with nowhere to go but down the conductive wire. Electrons traveling at a fast enough velocity along the confinement of a conductive wire are able to produce an electromagnetic field. The electromagnet field reveals the very nature of time itself which is a corkscrew event, wavelengths reveal this too. It is why special relativity can be associated with magnetics, electromagnetism, and a charge.

When a light is shined into a dark cave and all is revealed it does not mean the light created everything revealed in the cave, it only reveals the nature of the cave.

Because the nature of all things come from an infinitive state and the goal is to always return to an infinitive state somewhere in the structure of all things an infinitive state must exist.

 Infinity has to be able to adapt to all energies and velocities infinitively to mend all tears that can be created in infinity.

The wavelength, the corkscrew event is the infinite structure of all things created. The structure of each observable wavelength, corkscrew event is relative to the observer's mass, energy, and velocity.

Where a field of energy forms a line upon closer inspection if possible a corkscrew event will be revealed. Upon closer observation of the line of field produced to create this corkscrew event another corkscrew event would be observed and so on. All, only observable when observation is allowed and this relative to the observer's mass, energy, and velocity in accordance with the mass, energy, and velocity of what's being observed.

Faraday's law

With this all in mind it is no surprise for Faraday's law to exist. Faraday created a mocked up version of time itself, the corkscrew event. This is produced when a conductive wire is formed to produce a coil. A magnet is moved up and down in the coil. A magnet has already separated positive expanding time and dimensions and negative collapsing time and dimensions. Moving a magnet up and down in a mocked up version of the progression of time reveals time exists in strands and energy moving along the strand is relative.

Right Hand Rule of Thumb

The right hand rule of thumb; pointing your thumb in the direction of a current going down a conductive wire and your fingers curving in the direction of the electromagnetic field that will be produced. This is a very significant clue. This means the strands of time are tethered at one end because electromagnetic fields do not switch hemispheres when heading into the opposite direction. Time does not vanish behind us as we move into the future breaking the chain of events. Time appearing to vanish to the observer means the past is departing at or faster than our timeframe of reference which is the speed of light. It would be obvious to the observer which entity is capable of such a force as it would not emit light (black hole).

Electrons have enough qualities existing on their own in a newly forming universe to produce other forces needed to continue the evolution of matter.

Electrons have mass which means they have gravity. Electrons have expulsion qualities and a limit to how much force can be exerted upon them. Spinning together and orbiting under their own pressure in a newly forming universe they can create super accelerated magnetic fields. Like in a bubble chamber, electrons under pressure in magnetic fields can produce positrons an electron's antimatter.

In a newly forming universe electrons, under pressure start forming super accelerated magnetic fields, like sun spots. Each collapsing sphere of negative time and dimensions creates more pressure. This increases the rotation speed of the magnetic fields in the huge ball made of electrons, the mass equivalent of an average size galaxy. The rotation speed increasing creates enough centrifugal force to create looped magnetic fields. Surrounding electrons are pulled into these fields and accelerated out as positrons.

Positron

A positron is an electron's antimatter. The differences between the two are positrons are spinning in the opposite direction of an electron, and have a positive charge. This happens when more energy is applied to an electron and its energy field traveling on the outside sphere of negative collapsing time and dimension is pushed past the equator. This creates a spin in the opposite direction and a positive charge. It takes more energy to create a positron. Smashing a positron and an electron together can at times release enough energy to knock them back an evolutionary step into gamma rays.

Looped magnetic fields able to produce positrons, also create a Higgs field do to accelerated rotations and the further alteration of time.

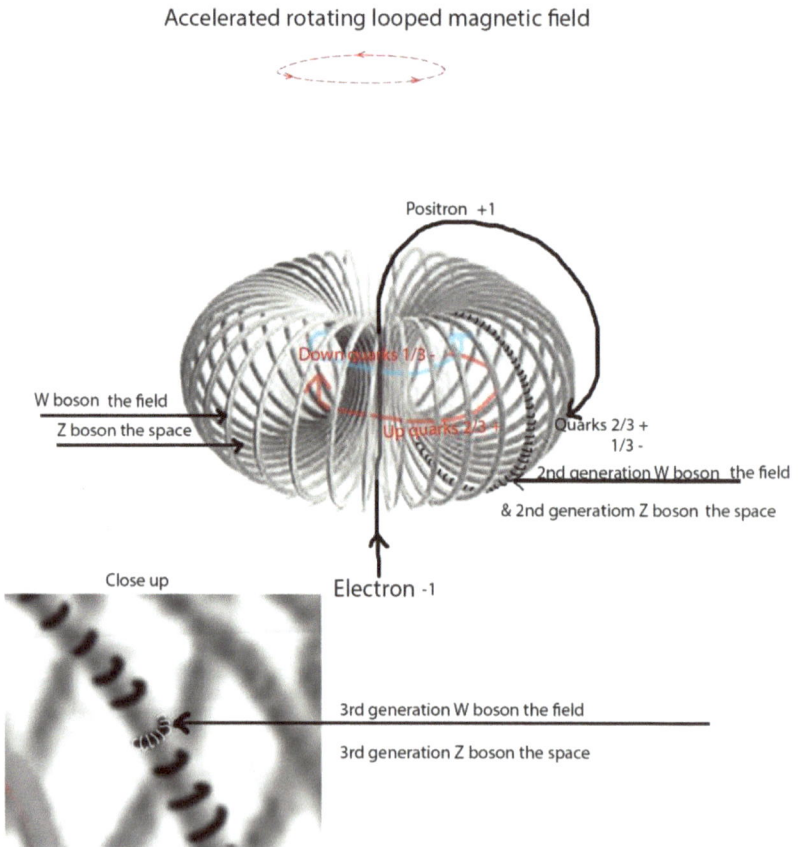

Accelerated rotating looped magnetic field

Positron +1

Down quarks 1/3

W boson the field
Z boson the space

Up quarks 2/3

Quarks 2/3 +
1/3 -

2nd generation W boson the field
& 2nd generatiom Z boson the space

Close up

Electron -1

3rd generation W boson the field

3rd generation Z boson the space

Bosons have one complete cycle of an interger spin compared to Fermions which have a half cycle interger spin. Fermions switch signs when positions are swapped bosons do not. Bosons can be in the same quantum state where fermions cannot.

Fermions with a half interger spin act like an electric motor. Its spin (Centrifugal force) creates a loop in a dimension in time at its equator, while separating positive and negative time and dimensions at the axis. The separation of time creates a magnetic north pole and a magnetic south pole at the axis. The loop created by the equator prolongs the effect of how time collapses and expands around it. Fermions have enough energy to circumnavigate on the inside of a collapsing sphere negative time and dimensions and on the outside. The change in the direction of time act like an alternating current in a loop able to spin in the freedom of space powered by each cycle. Fermions are able to filter time and its own energies through them by changing the size of their loop. Their loop is the amount of time and energy it takes to sustain their existence. Fermions act like free standing self-contained entities.

Bosons are created when a balancing act between collapsing negative time and dimensions and positive expanding time and dimensions is reached. A balance between these two opposing entities can occur on different levels, different generations and between fermions.

In a newly forming universe electrons are able to combine forces to create a more developed system able to build upon itself. An accelerated rotating looped magnetic field under immense pressure is able to create looped magnetic fields perpendicular to the looped magnetic fields created in the form of W bosons. All fields of energy are looped. The loops are created by the rotation of collapsing negative time and dimensions. The size and distance between the loops are relative to time, energy, mass, and velocity for both the loop and observer.

W Boson

A W boson loop acts like a loop of wire in a transformer. The difference is the wire in the transformer is in a fixed position and voltage can increase depending upon which side the loop is wound on the flux and the direction of energy supplied. When energy increases in a W boson loop system, more loops are created. Loops unwind as energy deceases. The fields that create the W boson loops are also looped creating an arena where different generations of quarks can combine. The different generations of W boson looped fields can produce different generations of space in the form of Z bosons.

Z boson

Z bosons are the spaces between the W boson's loops. The Z boson is just another generation of the Higgs Field. A Z boson is a balance created in the space between the W boson's fields. Different generations of the W boson's loops produce different densities of space creating the mass in the Z boson. Only the mass of the first generation Z boson is observed as it expels fermions in the form of neutrinos and an antineutrino (leptons) as it readjusts its density. Different generations of Z bosons expel different generations of neutrinos and antineutrino (electron neutrino, muon neutrino and tau neutrino). In different generations of the Z boson field different generations of an electron can form (electron, muon and Tau).

When the strands of time are cut or smashed false positives can be observed as time spirals in one direction into the future and spirals in another direction into the past.

In an infinite realm, if it is possible for the creation of a second generation it is possible for generations to progress infinitely. The observation of the generations of matter is also relative.

W and Z bosons create an arena allowing the formation and generations of quarks.

Quarks

Like everything else quarks need to stay within the chain of events from a system able to build upon itself over a period of time. No unexplainable entities allowed. It is well established in science if monopoles could exist it would be very, very rare or impossible. This does not appear to hold true when it comes to quarks. Quarks are excepted monopoles with fractionized charges in the world of science. All entities rotating at the right velocities have the ability to separate the effects of time. A magnetic moment is still a dipole and the moment is relative to the observer making it continuous. Entities creating a stabilized fractionized monopole effect must do so at the expense of their own physical attributes.

The chain of events are, infinitively connected spheres, wavelengths, electrons, W and Z boson fields, positron, quarks, gluons, Higgs boson, black hole and back to infinitively connected spheres.

Quarks are created from positrons when W and Z bosons create a Higgs field suitable enough for their existence. Quarks carry with them both two thirds positive and one third negative.

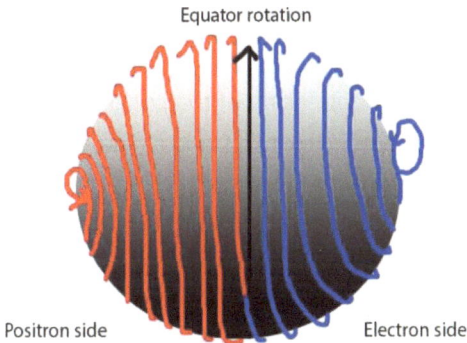

Equator rotation

Positron side   Electron side

This shows how an electron shifts into a positron. An electron changes its physical attributes and creates a monopole effect by turning a dipole inside out. Circumnavigating on the inside of a collapsing sphere of negative time and dimensions as positive energy and then circumnavigating along the outside as negative energy. The two opposing energies would cancel out if they were not separated by time. The whole system would fade out if it were not continually fed by collapsing negative time and dimensions.

Creation of quarks

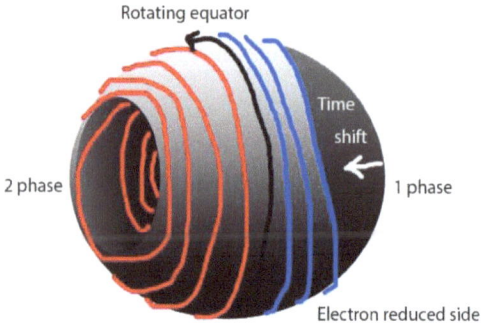

Rotating equator

Time
shift

2 phase

1 phase

Electron reduced side

Positron transformation to P particle

A positron changes into a monopole with a fractional charge by changing its physical attributes. As energy velocities increase a positron is able to circumnavigate again on the inside of the collapsing sphere of negative time and dimensions. The axis at this end of the sphere is where positive energy began as a wavelength. Positive energy lapping over itself only increases on the inside while fractionalizing on the outside. Because it is able to overlap one third of a positron's energy is equal to two thirds of the electron side of negative energy. Opposites attract, pulling two thirds of the electron side of the sphere inward.

Energy deficiencies can revert, the system back into an electron. Maximum energies and velocities can be reached as the positron's energy field reaches further into the sphere of collapsing negative time and dimensions. A maximum point is reached when the entire positive field enters the sphere.

The P particle elongates as opposing fields enter its core. The (P) in the P particle stands for Perfect particle.

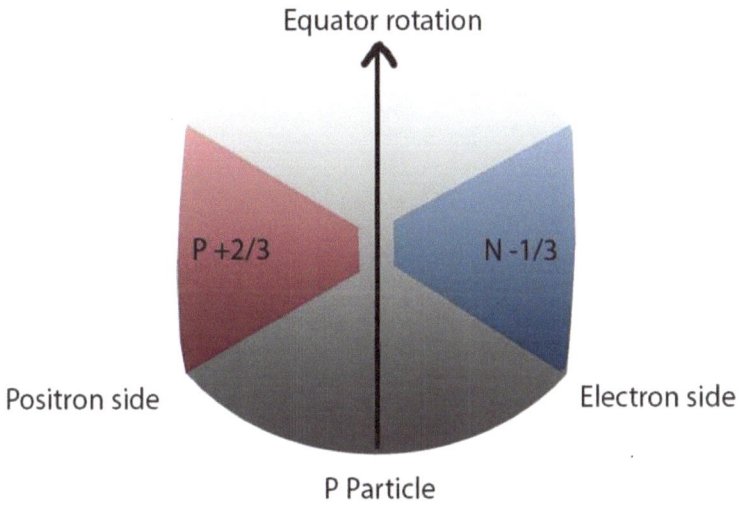

Equator rotation

P +2/3   N -1/3

Positron side   Electron side

P Particle

The P particle's ability to adjust its fields creates a more stable environment. The P particle's ability to fractionalize its charges can create a perfect environment for P particles to bond together.

Decay

$+3/3$ Positron → $-3/3$ Electron

Quarks

$+2/3$ — $-1/3$

$+1/2$ owe $-1/2$

$+1/3$ owe $-1/3$

0 Decay owe -1

$+2/3$ — $-1/3$

$+2/3$ — $-1/3$

$+1/2$ owe $-1/2$

$+1/3$ owe $-1/3$

$+1/2$ owe $-1/2$

0 Decay owe -1

$+1/2$ owe $-1/2$

$+1/2$ owe $-1/2$

$+1/3$ owe $-1/3$

$+1/3$ owe $-1/3$

$+1/3$ owe $-1/3$, 0 Decay owe -1

$+2/3$ owe $-1/3$

$+2/3$ owe $-1/3$

$+2/3$ owe $-1/3$

$+1/2$ owe $-1/2$

$+1/2$ owe $-1/2$

$+1/2$ owe $-1/2$

$+1/2$ owe $-1/2$

$+1/2$ owe $-1/2$

$+1/3$ owe $-1/3$

$+1/2$ owe $-1/2$

0 Decay owe -1

$+1/2$ owe $-1/2$

$+1/3$ owe $-1/3$

0 Decay owe -1

$+1/3$ owe $-1/3$

0 Decay owe -1

$+1/3$ owe $-1/3$

$+1/3$ owe $-1/3$

$+1/3$ owe $-1/3$

$+1/3$ owe $-1/3$

$+1/3$ owe $-1/3$

0 Decay owe -1

$+1/3$ owe $-1/3$

0 Decay owe -1

0 Decay owe -1

0 Decay owe -1

0 Decay owe -1

0 Decay owe -1

Rotary phase shift

W boson

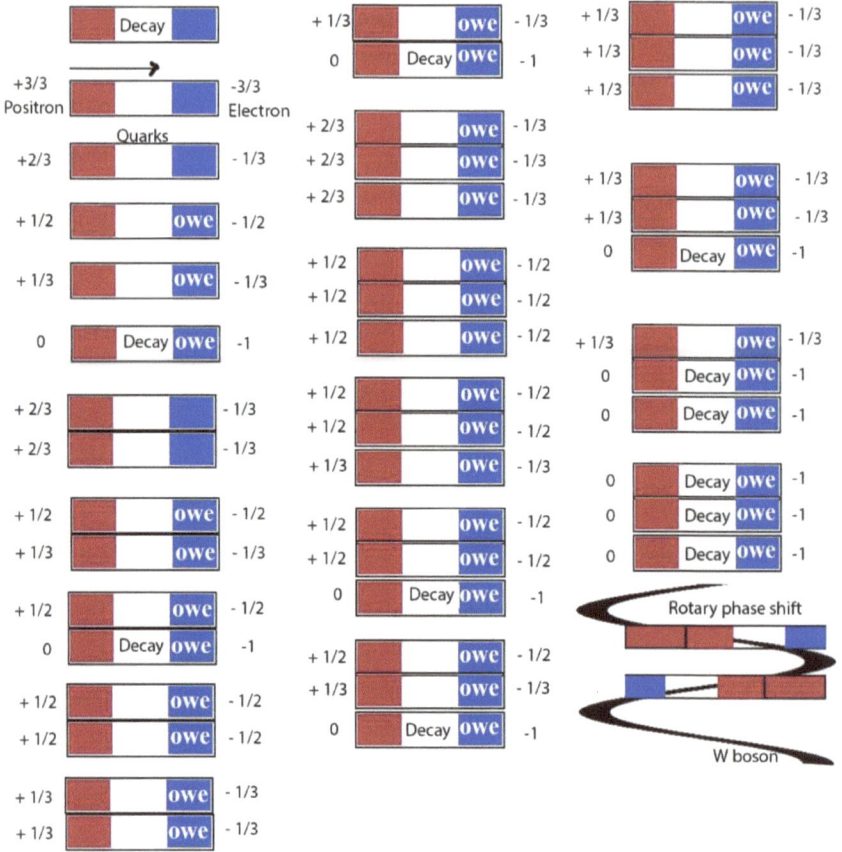

One third of fractionalized positive energy creates negative energy to fractionalize by two thirds. Positive energy fractionalizing by two thirds fractionalizes negative energy by four thirds. This leaves negative energy owing one third of its negative energy. Owing negative energy is another form of attraction making the system negative hungry. Negative hungry makes the system more electron hungry.

Looking at the chart it is easy to see it is hard to make one full positive charge without having an extra fractional negative charge or owing a negative charge.

The P particle is able to create two different kinds of quarks baryons (protons, neutrons) and mesons. Both are thought to be part of the hadrons family (grouped or paired particles). Baryons have a half interger spin making them fermions. Mesons have one interger spin making them bosons. Mesons are created when fermions are able to create a balance within the W and Z boson field. The balance is able to separate the axis and the equator and give the illusion of both particle and antiparticle when it is just a dipole. Mesons are able to create a balance in positive, negative, and neutral state (infinitive state). Mesons are created from the P particle.

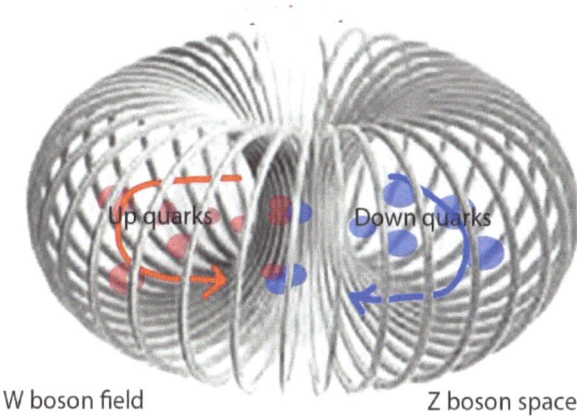

W boson field   Z boson space

Quarks opposing fields are relative to the motion of travel within the W boson looped magnetic field. The rotating W boson's fields of energy are plus or minus relative to the direction of the quarks as they travel through the Z boson space.

Positive expanding time

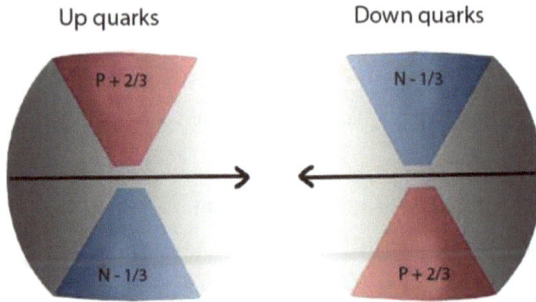

Up quarks   Down quarks

P + 2/3   N - 1/3

N - 1/3   P + 2/3

Collapsing negative time

Positive and negative charges produced by fermions are reversed depending on which dimension of time it favors.   This is not the case for bosons.

Positive time

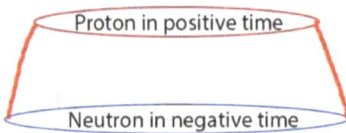

Proton in positive time   Neutron in positive time

Neutron in negative time   Proton in negative time

Positive energy moving into   Positive energy moving into
the future   the past

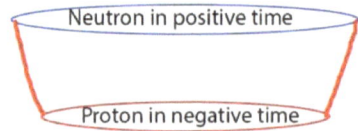

Negative time

Neutrons and protons revert into each other's state as time spirals both into negative and positive time. The connected ends of a strand of matter can connect as a proton, leaving the system with a negative one third surplus to compensate for. The end of a connected strand can also form into a neutron, leaving the system with a positive two thirds surplus to compensate for.

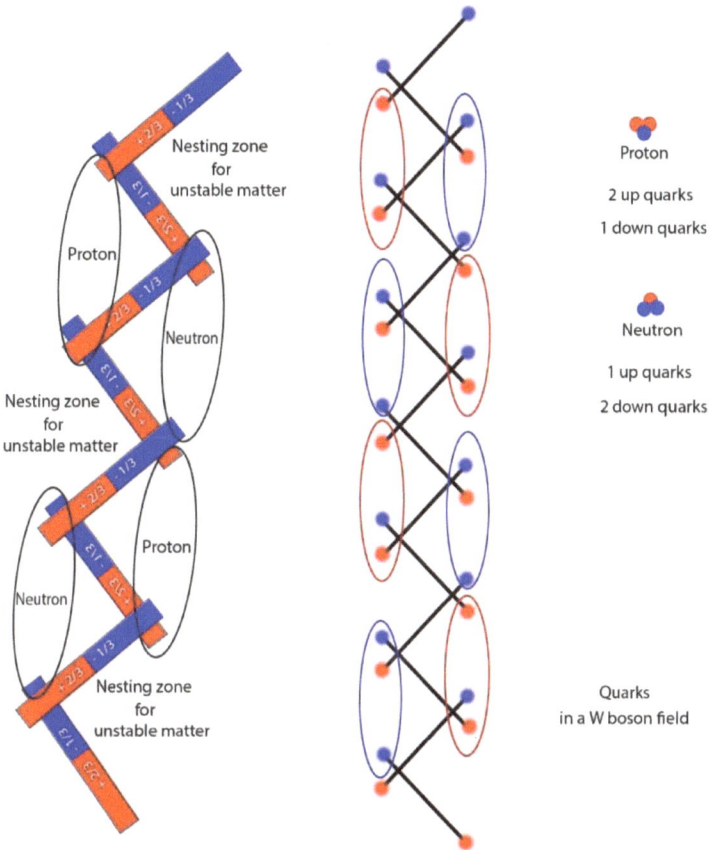

Mesons and hydrogen atoms can pull into nesting zones and make a strand of matter unstable by becoming neutrons or protons. This can create bonds with other systems as thin as a single electron.

The negative one third left over in denser matter is easily distributed between quarks and electrons making it hard to detect.

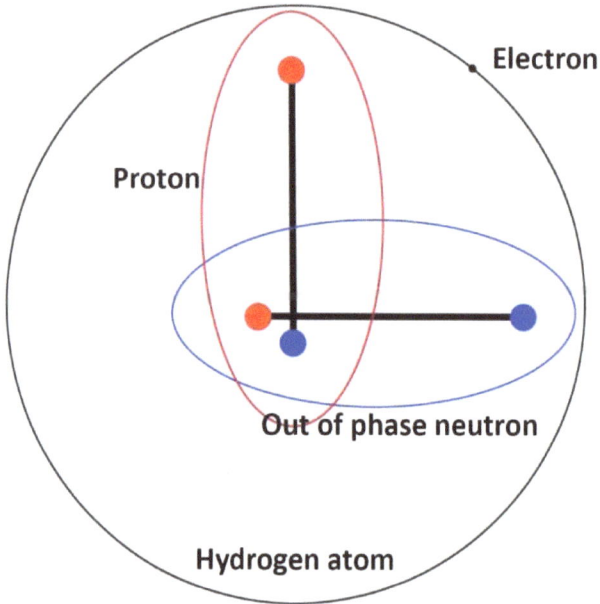

The negative one third is easy to detect in a hydrogen atom with one proton and one electron. In the hydrogen atom an electron is able to experience both states a proton and a neutron. The electron of three thirds negative charge is able to push, the one third negative side into a mesons state. This makes the hydrogen atom and its electron unique. The hydrogen atom is more bondable than any other atom and its electron slows down as it passes over the neutral state.

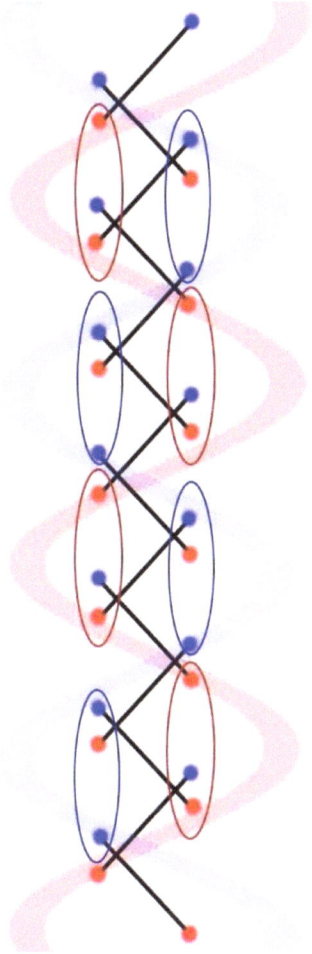

Time is able to move in opposing directions in a single rotation by neutrons becoming protons and protons becoming neutrons.

Gluons

In the Yan Yin Theory the God particle is not the Higgs Boson it is the Gluon and is more God like than just a gravity carrying particle.

Although gluons are considered to be massless they prove the entanglement of positive and negative time and dimensions is a strong force. The more entangled the stronger the force.

Gluons are created at connection points between the quarks in protons and neutrons and between protons and neutrons.

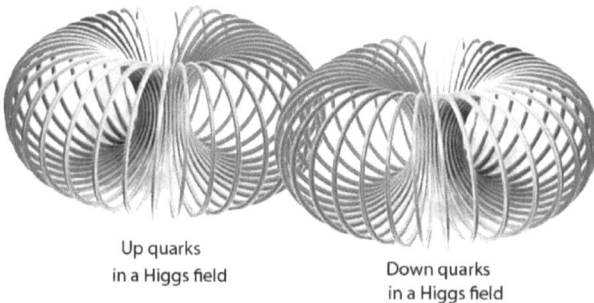

Up quarks
in a Higgs field

Down quarks
in a Higgs field

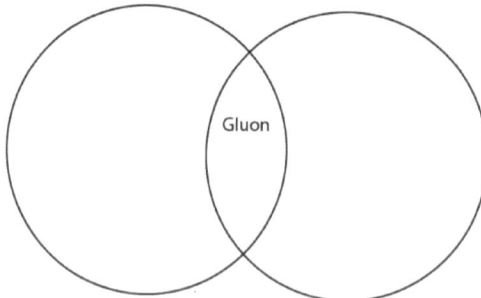

Gluon

The connection point between quarks acts like the gears in a clock. The difference is the teeth of the gear are loops (W bosons) and between the teeth is space (Z boson). When the gears turn they make a virtual balancing act in the form of a boson, the gluon. Gluons are unique because they are not created from the P particle instead they are created from the connection point between the P particles. The connection between quarks is called color confinement in the form of a color charge carried by the gluon.

Just like gears quarks can spin in different positions from one another. Because they are fractionalized dipoles two thirds positive on one end and one third negative at the other, connection points can come in many different combinations called flavors.

I did not mention the meson and anti-meson because they are a P particle with a balanced dipole.

Gluon in
green

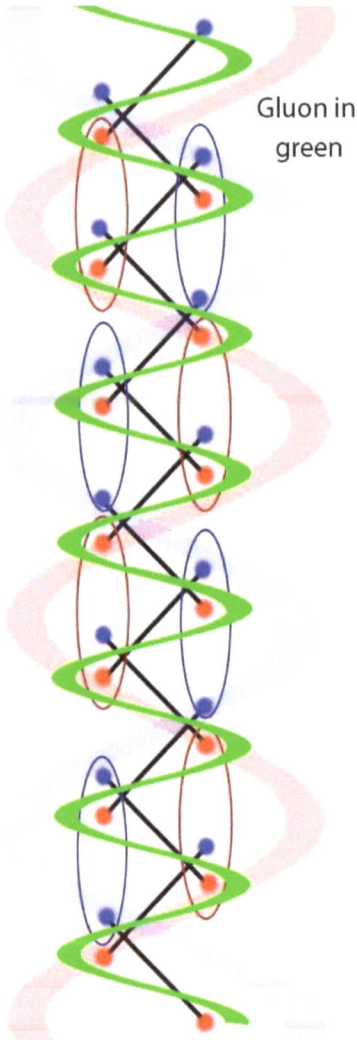

The gluon God factor is its ability to create P particles (quarks) when supplied with the right amount of energy in a process called hadronization. Hadronization is based on the same principles as breaking a magnet in half.

The connection points between the P particles do more than just create a virtual boson effect they create virtual time in the form of a virtual meson. A meson can create a balance in all three states, positive, negative, and infinite.

The gear system is made of up and down quarks pulling time in one direction and pushing it out in another. Having three gears it has the ability to merge or divide time. Through hadronization, divided time will become whole. Positive time and dimensions, cannot not escape the effects of collapsing negative time and dimensions on any level. Divided time put back together in a gluon tube will come out looking like two corkscrew events bypassing each other connected at ninety degree intervals representing the effect of the teeth (W bosons) in the gear system. In a more slightly sophisticated system based on the same principles it will produce a DNA strand.

Higgs Boson

In the Yang Yin Theory the Higgs Boson is a second generation gluon and called the black hole particle. It would take a huge amount of energy to observe this connection point.

Black holes are created when enough W and Z boson's looped magnetic fields connect together. The connection point creates a Higgs boson. Connected super accelerated looped magnetic field can create a runaway spiral staircase event creating a black hole.

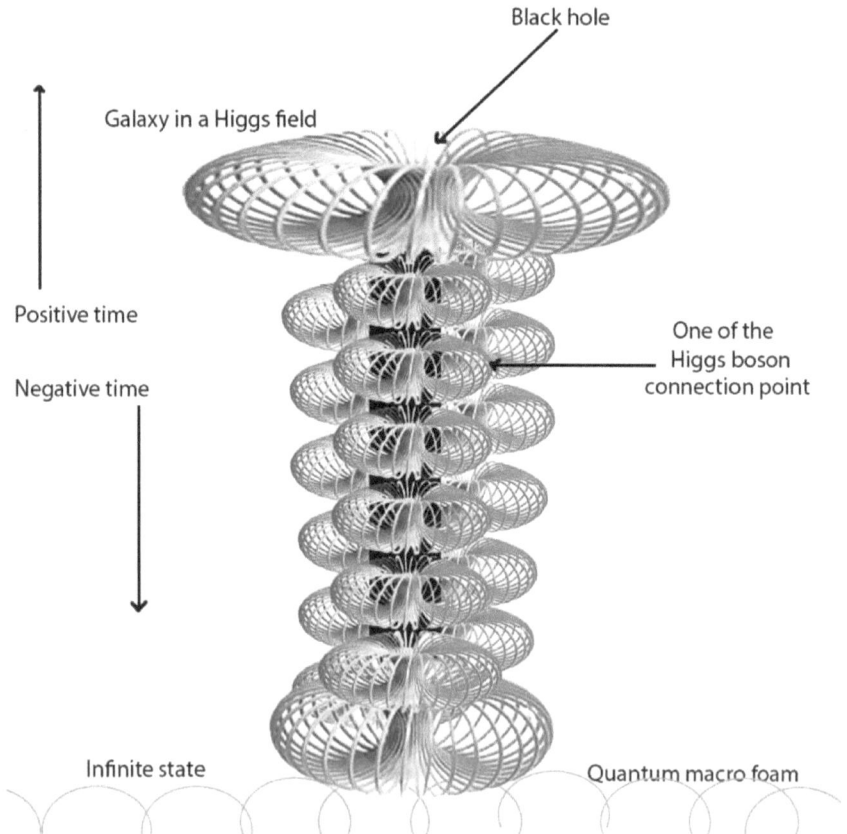

Black hole

Galaxy in a Higgs field

Positive time

Negative time

One of the Higgs boson connection point

Infinite state

Quantum macro foam

Galaxy

A galaxy is a self-contained entity held together by all its mass and its black hole in its center. A galaxy can exist perfectly on its own in a universe. If the big bang only created one galaxy it would still be a universe. A single galaxy in a universe may travel differently as the universe expands. A single galaxy may return to an infinite state at a different rate. A single galaxy and all its qualities, is all that needs to exist for a universe to exist. Our universe is a mega verse.

 Being a self-contained entity a galaxy has the ability to move into different latitudes on the inside of a collapsing sphere of negative time and dimensions according to its collective mass and velocity. This creates special relativity limiting what can be observed from within a galaxy.

Black Hole

A black hole acts like an adaptive immune system, trying to mend a tear in infinity by converting everything back into expanded quantum macro foam. For the most part black holes do just that.

It depends on how much junk matter is created in the process of making a black hole on how quickly the process will revert back into an infinite state. Junk matter is created when rotating looped magnetic fields (filled with bonding quarks) crash into each other and pull each other apart. This sends newly formed matter crashing into each other creating thermonuclear fusion. If enough matter is created in this process equaling the mass creating the black hole a balance is created in positive time, negative time, and an infinite state.

Black holes exist in most galaxies or will exist. Mass with a traveling velocity that exceeds the sphere of negative collapsing time and dimensions containment has the ability to create a black hole. The axis of a rotating collapsing sphere of negative collapsing time and dimensions are collapsing each moment from an infinite state. Mass traveling at certain velocities has the ability to travel to the highest possible latitude on the inside of a collapsing sphere of collapsing negative time and dimensions. The ability for mass to become a black hole is relative to its mass and velocity. The more volume and density a mass is taking up the less distance it needs to travel to reach a collapsing sphere of negative time and dimension's axis. The axis of rotating collapsing negative time and dimensions create a limit that matter can exceeded creating a black hole.

Being on the event horizon of a black hole physically is different than being on the event horizon of a black hole time wise. An object pulled into a black hole physically is pulled off the playing field. The object containing mass is pulled completely into negative time and dimensions through a black hole. Inside the black hole it is pilled like an onion, out of each of its atomic shells and expanded out, back into an infinitive state of EQMF.

Time wise a black hole has created a drain like effect. The way matter collapses into itself and expands outwardly is altered in the present of a black hole. Time is separated into past, present, and future. The present is pulled into negative collapsing time and dimensions revealing the future of each new moment in time. Positive energy has just enough energy to go into the next moment of time each moment while the past collapses individually into each atom and subatomic particle creating rotations and spins. The past also collapses as a whole, each celestial body spiraling back into and intertwining with the celestial body it is orbiting. The collapse of time, back into the past, is a three dimensional collapse, collapsing into one dimension dictated by the force contained in a black hole.

As time collapses from three dimensions to one dimension a rotation creating centrifugal force can create a prolonged two dimensional environment at the equator. This allows time to be filtered through all the matter existing in a two dimensional plane before it is pulled into one dimension and then into the past. Space is a dimensionless void where a three dimensional plane can exist in a two dimensional plane.

This is why most galaxies and most of the matter in the gravitational field of its black hole rotate like spokes in a wheel. It is three dimensions flattened into to two dimensions before the present is pulled into the past as one dimension into a black hole.

Black holes create their own special relativity within the galaxy and a universal special relativity. Time slows down as you near the event horizon of a black hole. As time changes so does observational perspective. As a black hole and all the matter in its gravitational field quantify at specific latitude in the sphere of collapsing negative time and dimensions it creates its own special relativity. Observable perspectives change at different latitudes in the sphere of collapsing negative collapsing time according to mass and velocities.

Every stage of the process of putting everything in a universe back into an infinite state exists at once. This is an infinitive observational point. This makes special relativity possible. Latitude and the mass a black hole quantified at and proximity to the event horizon of a block hole dictates what process the observer is in.

Before a black hole collapses in the creation of a galaxy/universe, it has a magnetic field, a magnetosphere. The majority of dense matter is metallic. A rotating galaxy equals a rotating black hole. The rotation of the majority of matter contained in the gravitational field of a black hole is not independent to the rotation of the black hole. Magnetism is an observable event relative to the observer. An infinite observation point of a black hole's magnetosphere cannot be observed in its entirety only its effect. A black hole's magnetic field is split up into a dipole in each dimension of positive and negative time. The effect is positive results in both dimensions while the core of the black hole exhibits the difference between the two fields which is zero. Gravitational and magnetic fields and the density of time and space help define the qualities of a black hole. A black hole helps to define the observable and physical moment.

The master key to the universe is not gravity, nor the speed of light, nor magnetism or electromagnetism, nor positive or negative time, nor strong nuclear force, its weak nuclear force.

Weak nuclear force

Weak nuclear force is a galactic event influenced by a universe generated event. The universe generated event is the expansion rate of the universe which can open up a Higgs field to the dimensionless void of space. The galactic event is the black hole's ability to close a Higgs field as it pulls it into negative time and dimensions. The structure of an atom can decay in either event. Depending on the mode of direction (positive or negative time) an atomic structure is favoring at the time of its decay will determine whether a W plus or W minus boson is revealed. The loss or closing or enlarging of a W boson field is the reason for the decay. The weakest point in a strand of matter is its nesting zone. The weakest link in an atomic structure is a neutron as it naturally wants to spiral back and forth into a proton and neutron state.

Single neutron decay

A singled out neutron will decay in just under fifteen minutes. This fifteen minute decay is very significant as it is influenced by a universe generated event, the expansion of the universe. This event has the ability to unravel a Higgs field. Tapping into this cycle is the key to mastering all things in this universe. Changing how time expands and collapses can change the influence of gravity. Tapping into this cycle can create the ability to slow down the rate of atomic and subatomic decay. Tapping into this cycle can slow down or speed up time.

It can also reverse the process of atomic and subatomic decay where matter can be reconstructed. If matter can be reconstructed on this level so can a DNA-strand. Life has already tapped into this cycle.

Creation of Life and beyond

If the creation of the universe were a computer program the formation of life would be a program glitch. The code that allows the formation of life is, a system able to build upon its self over a period of time with the ability to adapt to all energies and velocities exerted upon it. This is the same code for the formation of matter and a black hole.

Life, like everything else in a universe must create a balancing act between positive expanding time and dimensions and negative collapsing time and dimensions. The balancing act creating the existence for life must be continually fed for life to sustain itself in each new moment. This holds true for all that exists in a universe.

Creating a balancing act to sustain life must be unique or life would exist on every lunar and planetary system in a universe, asteroids included. The key is to find the unique balancing act, the Goldilocks zone or the CHZ.

Balancing points may only exist for moments. Black holes forming in a new galaxy could create a moment in time where past, present, and future exist at the same time creating an explosion of bacterial life. There could also be a zone with the same qualities approaching the event horizon of a black hole dumping bacterial life to freeze in space.

Atoms bond to form molecules. Molecules are connected spheres having their own unique vibrations. (Vibrations along the strands of infinitively connected spheres, is what creates a universe.) The vibrations molecules create are in the infrared zone. Molecules cease to move in zero-K, the temperature of space. If bacterial life is frozen in space a burst of energy might bring them back to life creating an infrared flux in space.

It is thought, life began in a different ocean other than what exist today, a primordial soup. This sounds like a good start because the salt in sea water. Also water, H2O is a unique molecule that has a high resistance in changing its own composition under its own weight. H2O has an electrical polar moment. Its boiling temperature changes under barometric pressure. This means the deeper the ocean the more energy it will take for it to boil. At the bottom of a deep ocean, salt water may have the right qualities to create a unique balancing act around a geothermal vent. I believe they have found single cell bacterial life in or around these vents. It has not been proven yet if this is where life began.

The inner planets in our solar system may have been gas planets including earth. They may have crossed paths intermixing atmospheres. This would have created huge electrical storms able to separate and create bonds between molecules producing a huge flow of positively charged ions and negatively charged anion. The gravitational pull between the planets may have created the unique balancing act needed for life to form.

A few reoccurring patterns stand out in life forms, the supply of continuous energy and its management, magnetism, positive, and negative charges, water, hydrogen, oxygen, and carbon. Carbon stands out because of its structure and bonding capabilities. Atomically it acts like a Higgs field, a boson. It acts like a gear system that is able to build upon itself as it incorporates other structures. The gear system is able to go beyond the internal life structures it supports and expand into the environment that supports the living structures.

The possibility for other intelligent life in our universe

The Drake equation attempts to answer this question. The world of science does not like this equation because there is a lot of guess work involved and there is missing variables. There have been such equations and theories presented throughout our history that needed to be adjusted, big bang included.

I would like to factor in the human hand and the evolution process it takes to create the human hand. I doubt the most intelligent mammal in the ocean with fins would ever communicate with other intelligent life in the universe. The creation of intelligence is an evolutionary process fueled by the struggle for survival.

In the evolutionary struggle for survival plants and animals are intertwined. Plants that can adapt to the environment and create a symbiotic relationship with its counterpart that uses it for fuel has a better chance for survival.

In the creation of the human hand plants would need to evolve into trees. Animals would have to feel the need to exist mainly in the canopy of the trees. The hand and the tail is the perfect solution in creating a better survival rate. Trees would then need to thin out or the animal pushed to where trees thin out for the evolutionary process to continue. It is why we don't see monkeys evolve into humans they are perfectly suited for their environment.

Walking upright and the use of hands a species can now evolve in the production of tools. The production of tools creates a new way to struggle to ensure survival. The production of tools is a way for a species to increase its intelligence without having to evolve physically.

Without the human hand or its equivalent intelligence would have to evolve along a different path. I doubt that an alternative path would lead to extraterrestrial communicating by conventional means. Without the human hand or its equivalent space travel would be doubtful as well. The human hand or its equivalent is crucial in creating a path for higher intelligence to evolve.

In terms of Infinity if it can happen once, it can happen, an infinite amount of times. The problem is an evolving universe is in a finite state. The question is; is our universe vast enough to recreate the incredible path of evolution it took to create us?

The answer maybe yes, and we are not listening or observing correctly.

The speed of light our holy grail may not be constant, which is my belief. It is just the latitude on the inside of collapsing negative time and dimensions our galaxy quantified in. It may or may not be at the equator of collapsing negative time and dimensions (universal time).

The point is a thing called gravitational time dilation and frequency shift. From galaxy to galaxy frequencies may expand or contract depending on the galaxies total mass and the strength of its black hole. It may be a simple matter of filtering frequencies as they come in to match our timeframe of existence.

Another possibility is extraterrestrial beings are using a more advanced communication system that we view as a natural phenomenon. If extraterrestrials were communicating with each other across the vastness of space and they had the ability they would choose the fastest means possible. If they could communicate on a daily basis they would. The only observable entity so far that fits this scenario is a gamma ray burst. The code of information could be written in the elements exploded. Gamma ray bursts could also be a form of universal warfare or a faster means of travel. The simple answer for a gamma ray burst is a supernatural event.

The possibility of reaching and inhabiting extraterrestrials planets or moons

We have already done the first part by landing on our own moon. I am very disappointed in NASA for not creating a base on the moon as of yet. The problem is funding, which becomes political, and politics is logic squeezed until something stupid comes out. The public lost interest in our moon so NASA because of funding moved its focus towards Mars.

It is crucial for us as a species to seed the universe with human life forms. It is crucial for us to create a base on the moon. On moons with less gravity than the Earth we humans appear to have superhuman strength. We can carry more. Spacesuits need to be made with this in mind. It is easier to lunch a spacecraft where there is less gravity. Solar panels fueled by direct sunlight on our moon could create a laser system able to protect the Earth from asteroids. While on the subject of protecting the Earth, I suggest putting an electromagnetic making device in the Earth's lagrangian 1 point to combat coronal mass ejections produced by our sun. This device could also help to regulate the Earth's temperature.

I'm glad to see the interest in Phobos one of Mars's moons. In the future when capabilities allows, moving Phobos in a tighter orbit around Mars might help in re-liquefying Mars's core. This would create a magnetosphere that would help to sustain a livable atmosphere and a more suitable temperature.

While were moving moons around, it might be in our best interest to move Deimos (Mars's other moon) in an orbit that is closer to earth. This would make a much needed resting and resupplying point between our two planets. This might also help in liquefying Mars's core as the gravity of the two moons might make the core of Mars too stable.

It is very doubtful we are going to travel through the vastness of space in spacecraft that are now typically designed or those imagined by Hollywood. A typical spacecraft traversing the universe would take on too many hits from space debris.

It just so happens we are in luck. Our future space crafts are buzzing around our solar system in the form of asteroids. We will be travelling the vastness of space in spaceships that look more like giant potatoes than the Starship Enterprise.

Burrowing into an asteroid would shield us from radiation and impacts. Growing food and the production of materials would be much easier. Unstable asteroids could be drilled in the correct locations where appropriate material would be placed. The unstable asteroid would be moved into a close orbit with the sun to heat and temper it. Leaving our solar system would encompass a trip to a few more moons and to the rings of Saturn to pick up an asteroid of ice.

I do not see us making vast trips into space alone. We will need to take robots that look like us with nurturing qualities. These robots will need to be sophisticated enough to recreate life from test tubes. Food supplies and oxygen production will dictate the crew member size. Food and oxygen supplies will need to build up to supply crews large enough to have their own offspring to continually supply the gene pool with new genes.

Robots will need to know every aspect of running and maintaining the craft and themselves. Intelligent levels amongst humans may not reach a high enough level to run and maintain every aspect needed to proceed continuously into the vastness of space. Because intelligent levels are a hit or miss with humans, and can change with age, robots will need to be in charge. Robots will relinquish their authority momentarily when humans show the correct aptitude to take charge. Mistakes made in the vastness of space in a small craft are unforgiving.

It will be a matter of, how long can our genes maintain their integrity in a frozen state, if or when a robot will create us in a test tube before reaching our destination.

The possibility of an ultimate supreme being (God)

Science and Gods do not mix. Fanatics can exist in both arenas of science and religion.

Science is the quest for the ultimate truth. A "truth" that must go through a scrutiny of tests before it is labeled a fact.

Religion is a quest to gather followers who are willing to give faith to a popular excepted belief. A belief does not have to be a proven fact and is why there are so many different churches and there have been so many concepts of God or Gods.

In my theory for the present time I have many unproven facts as do all theories relating to the origin of the universe. The difference is the Yang Yin theory has an unbroken chain of events from start to finish and all entities and patterns used, exist in this universe.

I believe I would be taken more seriously leaving the God theory out and would if it had not presented itself in one of the patterns created in the creation of the universe.

The funny thing is I am trying to create a revolution in science somewhat like Copernicus both needing a Kepler. Copernicus felt the need to apologize to religion whereas I feel the need to apologize to science. The difference is I am not on my death bed as of yet, sorry, Fanatics.

The God pattern created in the Yang Yin Theory is universes continually being created and put to rest in the endless sea of expanded quantum macro foam. The micro entity that resembles this macro process the best is, living cells, as they look like expanded quantum macro foam, and the creation of a universe, best resembles the thought process created by free thinking entities (synapse). Created in its image, may mean something more scientific encompassing all free thinking entities.

Expanded quantum macro foam through vibrations is able to create universes that mainly follow the laws of physics without creating a free thinking entity. One the same lines a free thinking entity does not have to think about all the physics involved in order for it to think or maintain its body.

The infinitively connected spheres of EQMF thought process maybe completely depended upon the free thinking entities it is able to create. This would make the evolution of free thinking entities and its evolution the same process where "I am what you say I am" a true statement.

Because EQMF is in an infinite state, it's been creating free thinking entities endlessly. The answer to the question; who thought of who first, would be the same moment because moments don't exist in an infinite state. Moments only exist in the minds of free thinking entities.

The bottom line is science may never find a way to prove how exactly the universe began or how it will end or how to create a single cell living entity from a chemical reaction. The close as science may get is a popular excepted scientific guess.

Religion is more interested in faith than establishing facts. The concept of God and heaven is not of this world, where facts do not apply only the imagination of free thinking entities. If the connections between EQMF and free thinking entities are true imagination might be enough. Imagination and reality is as intertwined as the evolution of plants and animals.

Science and religion both believe, as do I, only time will tell.

Will the human species survive the test of time?

As a species sharing a planet with other species we have broken the law of natural selection through our technological advancements. Our goal is anybody and everybody can survive and as long as humanly and mechanically possible. This is all without regard to the Earth's limited amount of natural resources.

China is the only modern country that has tried to regulate its population.

We have become out of balance with our ecological system. We have become a parasite and will die like a virus unchecked by killing our host the ecosystem.

We may have already crossed the window of opportunity and are heading for a catastrophic event. It is something that should not be taken lightly, everything we know and love is at risk.

The sacrifices that need to be made are going to keep increasing as time flows by. This means sacrifices are going to become harsher as politics are squeezed and politicians feel the need to survive.

The best solution with the least amount of throwaway people is to seed the universe. Most societies throughout time have produced throwaway people. Throwaway people are those who do not contribute to a society but draw from it. This does not include the young with parental support or the retirees who have a prepared or prepared for their retirement. Most revolutions begin when throwaway people become forty to forty five percent of a society's population. When weaponry outweighs building and farming equipment the path to more throwaway people has begun. Every life is important when your intent is to seed the universe with life.

The other solution is to follow suit with China and regulate the world population globally. On a smaller scale it is happening anyways through war and famine. The modern wars we are in and the wars to come are predictable. The civilizations that are least technologically advanced held back in the quagmire of tradition and religion are sitting on the most unused natural resources. Because we are a globally divided species, when the time comes, wiping each other out will be easier than if we were globally united.

Another solution is to create a league of scientists who will lay out a scientific path to insure our survival that all civilizations must adhere to and enforced by a true world army. People with controlling power usually never relinquish their control until they are forced to do so.

Economics is the new struggle for survival. Concerned governments, agencies, and citizens manage the best they can and struggle for those who cannot.

I'm hoping my theory about weak nuclear force is correct and will act as a catalyst to insure the survival of our species.

Summary

Nothing from nothing is, always nothing. If this statement is not true, you are most likely leaving out something or adding something.

All finite entities have a clear chain of events in the evolution of its creation. All the links in the chain of the events may or may not be observable in a finite realm.

Infinity is the only entity that can exist before time and space. To think otherwise is breaking the chain of events, where links of events can appear and disappear infinitely allowing no direction in time, where chain of events are meaningless.

The quest for the truth is the quest to our existence. Once answered a bucket of water will still be a bucket of water and a stick still a stick for those who only know them as such. Those who are unchanged are those who would be left in the dark if it were not for science.

Once the truth to our existence is found a computer program will be written, where all will have access to the evolution, to a countless amount of universes. It may take hundreds of years with millions of programs running to create one with life or it may happen with the first correctly written program. Once one program produces life all other programs will be updated.

In a universe generated by a computer program, space travel simulation will produce real results. Simulated scientific experiments will give real results. If and when a computer generated program produces intelligent life they may find answers to problems before we do.

Science is always looking for the magic equation that will answer all things. The problem is you have to know it when you see it. Energy equaling mass times the speed of light squared maybe the right equation if it is taken literally and in the right context.

The speed of light means much more when realized it is the timeframe of our reality. It is why it is the fastest rate that can be measured. Light is constant because it is able to stretch across each frame of time in every frame. Like drawing a line across each, frame in every frame in a movie picture reel. Light just happens to travel at the same speed as the timeframe of our reality.

Taking the constant speed of light in this context means the constant speed of light times itself must be taken literally. This means if matter were taken apart and stretched out in photons it would stretch out to a huge astronomical distance in one second. Just putting the constant speed of light squared back, as photons in the confinement given by mass cannot be done as a solid without creating a density problem. It's why loops are the perfect answer to the solution. Mass equals energy divided by the constant speed of light squared is literal. A timeframe and the size of its space, is relative. Energy is a looped event divided by frames of time (the constant speed of light).

The faster a mass travels the slower its personal clock becomes. You have to know when you are looking at something that is traveling so fast it is traveling slower. That's what atomic and subatomic particles are doing. We the observers are living in the micro world with huge vast amounts of space. The macro world is the atomic and subatomic world where energy has taken up all the space and time allowed by a universe dictated by a never ending collapsing negative force of negative collapsing time and dimensions.

Combining the equation of quantum mechanics and general relativity proves the Yang Yin Theory is correct as the end results is infinity of endless loops.

I did not write this theory for scientific or religious fanatics. I wrote this theory for those of you are willing to wake up to a different dawn. To those of you who are willing to push the limit in the quest for truth where politics have no meaning.

Conclusion

The Yang Yin Theory will be labeled as conjecture and the ravings of a crackpot. The science world likes the word "crackpot". Kinder views will say things like, "pseudoscience", and "hypothetical".

I am OK with any and all conclusions made by all who fully read and fully understand this theory.

51

I believe I created a theory that bridges the gap between science and religion.

Science has too many inconsistencies, and too many unanswered questions opening the door to religious conjecture. On the other hand, Science is always willing to improve upon itself.

The science of religion is intertwined with its history and the free thinking entities feeling the need for its existence.

Not being in control of one's own environment is where the need for religion stems.

Lock a person up in solitary confinement for a long period of time and they will, kill themselves, or go insane, or find religion. This is taking away a person's control over their environment to the extreme.

A person on their death bed is another extreme, where many nonbelievers convert.

Drug abuse, is a person continually dumping their control over their own environment into an empty void of self-destruction. Their drug of choice at times can be switched to religion. The drug use may have started because they were not in control over their own environment.

More in the past, but still prevalent today supernatural events (hurricanes, tornados, earthquakes, volcanoes, meteorites), could make people feel they are not in control over their environment. Creating a father figure that would look over them in their times of need would be most reassuring.

There are too many people involved with this way of thinking for it to be taken lightly. I commend those who take part in religion for the advancement of themselves and the advancement of the human race.

History paints another side of religion and thus a lesson to be learned. Religion like science also has too many inconsistencies, and too many unanswered questions. Religion needs to learn it is OK to improve upon itself as new facts present themselves the way of science.

Religion must admit it has done and still is involved in horrific acts against humanity. It does not matter if it is done by extremists if it is in the name of religion, history is unforgiving.

My point is, religion like science has evolved over time but at a much slower pace. Religion has evolved from many Gods, to one God, and to a son of God. Religion has moved so much slower than science it is two thousand years behind. Two thousand years ago, science wise we were like children in how we viewed our environment. We thought the world was flat and in the center of the universe, if religion had its way that's where we would be today.

I myself would not invest my entire existence on the mindset of a child. I would hope as we leave this Earth to seed the universe we do so as adults who are in charge of their own environment.

It is not the sign of a false prophet but one in an adolescent stage of awareness when God is referred to as Father or he. Religion was conceived from a male dominated world where females were considered to be second class citizens. If we were cattle a bull would be the image of our God.

In this way we are a very arrogant species. We attach a dominate image of ourselves to feel more connected thus more secure.

Scientifically the concept of a single supreme being having one or any gender that relates to us does not make sense. The evolution of gender is created as a species becomes more evolved and infant development consumes more time.

The world of science has fallen into the same male dominated egocentrics ideology. Concerned only with a singularity and its ability to seed the universe with matter has left science blind to the womb and its forces housing a singularity.

To help bridge the gap between religion and science both need to come to terms with their weaknesses, both need to evolve.

Our primary goal as a species should be to seed the universe with life. Science, government, and religion all need to make this their primary objective. Over populating the planet becomes a positive only when this objective is achieved.

Evolving over a period of time is the natural order of things. Not to evolve can be unnatural. We have had our Mayan 25 thousand year cycle. It is time to move on. I predict in the next 25 thousand years we will occupy five other planets in this galaxy and one planet in another galaxy. The universe is our next frontier the time for waiting is over. The time to evolve is now.

54

I have been working on this theory for over thirty years. Most of my time has been wasted stuck in a box the world of science has created. I had an epiphany in my sleep and realized we mere humans usually get things completely backwards before we get them right. This is when I realized the missing part to the puzzle is negative collapsing time and dimensions.

I apologize for the lack in my writing skills.
(In the voice of Forest Gump) I am not a smart man but I know what the universe is.

Any question or comments please go to

Yangyintheory.com

Reference

Albert Einstein

Max Planck

Michael Faraday

James Maxwell

Sir Isaac Newton

Johannes Kepler

Nicolaus Copernicus

Nikola Tesla

Galileo Galilei

Fred Hole

Theodor Kaluza

Karl Schwarzschild

Subrahmanyam Chandrasekhar

Wolfgang Pauli

Murray Gell-Mann

George Zweig

Chin Ning Yang

Robert Mills

Santiago Ramon y Cajal

Paul Winterbert

www.ingramcontent.com/pod-product-compliance
Lightning Source LLC
Chambersburg PA
CBHW041715200326
41519CB00001B/168